FRACTIONS WORKBOOK

Grade 5 Math Essentials

Children's Fraction Books

BABY PROFESSOR

EDUCATION KIDS

Have fun and enjoy solving fraction problems with this workbook.

Equivalent Fractions

Fill in the missing numerators and denominators for the equivalent fractions.

1. $\dfrac{5}{20} = \dfrac{}{24} = \dfrac{}{8} = \dfrac{}{4} = \dfrac{3}{} = \dfrac{4}{} = \dfrac{7}{}$

2. $\dfrac{6}{12} = \dfrac{}{2} = \dfrac{3}{} = \dfrac{2}{} = \dfrac{7}{} = \dfrac{}{8} = \dfrac{5}{}$

3. $\dfrac{6}{15} = \dfrac{}{20} = \dfrac{12}{} = \dfrac{}{10} = \dfrac{2}{} = \dfrac{}{35} = \dfrac{}{25}$

4. $\dfrac{6}{16} = \dfrac{}{48} = \dfrac{}{40} = \dfrac{21}{} = \dfrac{}{32} = \dfrac{9}{} = \dfrac{}{8}$

5. $\dfrac{6}{7} = \dfrac{36}{} = \dfrac{42}{} = \dfrac{}{35} = \dfrac{12}{} = \dfrac{}{21} = \dfrac{}{28}$

1. $\dfrac{12}{18} = \dfrac{}{6} = \dfrac{2}{} = \dfrac{}{12} = \dfrac{10}{21} = \dfrac{}{} = \dfrac{6}{}$

2. $\dfrac{54}{60} = \dfrac{45}{} = \dfrac{36}{} = \dfrac{63}{} = \dfrac{9}{} = \dfrac{27}{} = \dfrac{18}{}$

3. $\dfrac{7}{42} = \dfrac{4}{} = \dfrac{}{36} = \dfrac{}{12} = \dfrac{}{6} = \dfrac{5}{} = \dfrac{}{18}$

4. $\dfrac{4}{9} = \dfrac{}{54} = \dfrac{20}{} = \dfrac{}{63} = \dfrac{}{36} = \dfrac{8}{} = \dfrac{12}{}$

5. $\dfrac{7}{49} = \dfrac{1}{} = \dfrac{3}{} = \dfrac{}{14} = \dfrac{}{28} = \dfrac{6}{} = \dfrac{5}{}$

1. $\dfrac{4}{9} = \dfrac{8}{} = \dfrac{12}{} = \dfrac{}{36} = \dfrac{20}{} = \dfrac{24}{} = \dfrac{28}{}$

2. $\dfrac{3}{4} = \dfrac{}{8} = \dfrac{}{12} = \dfrac{}{16} = \dfrac{}{20} = \dfrac{}{24} = \dfrac{21}{}$

3. $\dfrac{5}{8} = \dfrac{}{16} = \dfrac{15}{} = \dfrac{}{32} = \dfrac{}{40} = \dfrac{}{48} = \dfrac{}{56}$

4. $\dfrac{3}{5} = \dfrac{6}{} = \dfrac{}{15} = \dfrac{}{20} = \dfrac{15}{} = \dfrac{}{30} = \dfrac{}{35}$

5. $\dfrac{2}{7} = \dfrac{}{14} = \dfrac{}{21} = \dfrac{8}{} = \dfrac{}{35} = \dfrac{12}{} = \dfrac{14}{}$

1. $\dfrac{2}{7} = \dfrac{4}{} = \dfrac{}{21} = \dfrac{}{28} = \dfrac{10}{} = \dfrac{12}{} = \dfrac{}{49}$

2. $\dfrac{1}{2} = \dfrac{2}{} = \dfrac{3}{} = \dfrac{4}{} = \dfrac{5}{} = \dfrac{}{12} = \dfrac{7}{}$

3. $\dfrac{5}{6} = \dfrac{10}{} = \dfrac{}{18} = \dfrac{}{24} = \dfrac{25}{} = \dfrac{}{36} = \dfrac{}{42}$

4. $\dfrac{1}{10} = \dfrac{}{20} = \dfrac{}{30} = \dfrac{4}{} = \dfrac{}{50} = \dfrac{}{60} = \dfrac{}{70}$

5. $\dfrac{1}{3} = \dfrac{}{6} = \dfrac{3}{} = \dfrac{}{12} = \dfrac{}{15} = \dfrac{6}{} = \dfrac{}{21}$

1. $\dfrac{4}{9} = \dfrac{8}{} = \dfrac{}{27} = \dfrac{16}{} = \dfrac{20}{} = \dfrac{}{54} = \dfrac{28}{}$

2. $\dfrac{1}{3} = \dfrac{}{6} = \dfrac{3}{} = \dfrac{4}{} = \dfrac{5}{} = \dfrac{}{18} = \dfrac{7}{}$

3. $\dfrac{4}{5} = \dfrac{}{10} = \dfrac{}{15} = \dfrac{}{20} = \dfrac{}{25} = \dfrac{}{30} = \dfrac{}{35}$

4. $\dfrac{1}{7} = \dfrac{2}{} = \dfrac{3}{} = \dfrac{}{28} = \dfrac{}{35} = \dfrac{}{42} = \dfrac{7}{}$

5. $\dfrac{1}{2} = \dfrac{}{4} = \dfrac{}{6} = \dfrac{4}{} = \dfrac{}{10} = \dfrac{}{12} = \dfrac{}{14}$

1. $\dfrac{7}{10} = \dfrac{}{20} = \dfrac{21}{} = \dfrac{}{40} = \dfrac{35}{} = \dfrac{42}{} = \dfrac{}{70}$

2. $\dfrac{1}{6} = \dfrac{}{12} = \dfrac{3}{} = \dfrac{}{24} = \dfrac{5}{} = \dfrac{}{36} = \dfrac{}{42}$

3. $\dfrac{3}{4} = \dfrac{}{8} = \dfrac{}{12} = \dfrac{12}{} = \dfrac{15}{24} = \dfrac{}{24} = \dfrac{21}{}$

4. $\dfrac{4}{7} = \dfrac{}{14} = \dfrac{}{21} = \dfrac{16}{} = \dfrac{}{35} = \dfrac{24}{} = \dfrac{28}{}$

5. $\dfrac{1}{8} = \dfrac{2}{} = \dfrac{}{24} = \dfrac{4}{} = \dfrac{}{40} = \dfrac{}{48} = \dfrac{}{56}$

Equivalent Fraction Problems

Fill in the missing numerators and denominators for the equivalent fractions.

1. $\dfrac{1}{2} = \dfrac{}{12}$

2. $\dfrac{6}{15} = \dfrac{}{5}$

3. $\dfrac{2}{3} = \dfrac{}{6}$

4. $\dfrac{}{4} = \dfrac{8}{16}$

5. $\dfrac{}{18} = \dfrac{5}{6}$

1. $\dfrac{}{10} = \dfrac{3}{5}$

2. $\dfrac{24}{} = \dfrac{4}{5}$

3. $\dfrac{}{25} = \dfrac{4}{5}$

4. $\dfrac{5}{} = \dfrac{1}{2}$

5. $\dfrac{3}{4} = \dfrac{}{16}$

1. $\dfrac{1}{5} = \dfrac{5}{}$

2. $\dfrac{2}{} = \dfrac{1}{2}$

3. $\dfrac{6}{24} = \dfrac{}{4}$

4. $\dfrac{5}{} = \dfrac{1}{6}$

5. $\dfrac{5}{6} = \dfrac{10}{}$

1. $\dfrac{}{10} = \dfrac{1}{2}$

2. $\dfrac{}{8} = \dfrac{2}{4}$

3. $\dfrac{1}{5} = \dfrac{}{10}$

4. $\dfrac{}{5} = \dfrac{12}{15}$

5. $\dfrac{6}{10} = \dfrac{}{5}$

1. $\dfrac{1}{2} = \dfrac{4}{\_}$

2. $\dfrac{15}{\_} = \dfrac{3}{4}$

3. $\dfrac{18}{\_} = \dfrac{3}{4}$

4. $\dfrac{18}{36} = \dfrac{3}{\_}$

5. $\dfrac{\_}{6} = \dfrac{2}{3}$

1. $\dfrac{18}{} = \dfrac{3}{4}$

2. $\dfrac{1}{5} = \dfrac{5}{}$

3. $\dfrac{}{8} = \dfrac{1}{4}$

4. $\dfrac{6}{12} = \dfrac{}{2}$

5. $\dfrac{18}{36} = \dfrac{}{6}$

Mixed to Improper Number

Convert the Mixed to Improper Number.

1. $9\frac{3}{4}$ = _____

2. $7\frac{2}{7}$ = _____

3. $7\frac{2}{5}$ = _____

4. $8\frac{2}{3}$ = _____

5. $7\frac{1}{2}$ = _____

1. $7\frac{7}{10} = $ ____

2. $5\frac{1}{2} = $ ____

3. $7\frac{1}{3} = $ ____

4. $6\frac{5}{9} = $ ____

5. $7\frac{4}{5} = $ ____

1. $2\dfrac{1}{8}$ = ____

2. $4\dfrac{5}{6}$ = ____

3. $8\dfrac{1}{3}$ = ____

4. $7\dfrac{5}{7}$ = ____

5. $9\dfrac{7}{9}$ = ____

1. $8\frac{1}{4}$ = _____

2. $7\frac{2}{5}$ = _____

3. $2\frac{3}{7}$ = _____

4. $7\frac{1}{5}$ = _____

5. $6\frac{1}{2}$ = _____

1. $4\dfrac{1}{5}$ = _____

2. $4\dfrac{1}{2}$ = _____

3. $2\dfrac{3}{4}$ = _____

4. $5\dfrac{2}{3}$ = _____

5. $9\dfrac{4}{5}$ = _____

1. $6\frac{7}{10}$ = ____

2. $6\frac{2}{5}$ = ____

3. $3\frac{4}{5}$ = ____

4. $6\frac{1}{2}$ = ____

5. $2\frac{5}{9}$ =

1. $3\dfrac{5}{8} =$ ____

2. $9\dfrac{1}{2} =$ ____

3. $3\dfrac{1}{2} =$ ____

4. $4\dfrac{6}{7} =$ ____

5. $9\dfrac{1}{5} =$ ____

Improper to Mixed Number

Convert the Improper to Mixed Number.

1. $\dfrac{27}{4} = $ _____

2. $\dfrac{46}{8} = $ _____

3. $\dfrac{45}{10} = $ _____

4. $\dfrac{39}{10} = $ _____

5. $\dfrac{32}{7} = $ _____

1. $\dfrac{21}{9} = $ _____

2. $\dfrac{13}{3} = $ _____

3. $\dfrac{37}{8} = $ _____

4. $\dfrac{13}{2} = $ _____

5. $\dfrac{29}{4} = $ _____

1. $\dfrac{31}{4}$ = _____

2. $\dfrac{14}{4}$ = _____

3. $\dfrac{12}{5}$ = _____

4. $\dfrac{5}{2}$ = _____

5. $\dfrac{14}{5}$ = _____

1. $\dfrac{10}{3}$ = _____

2. $\dfrac{38}{5}$ = _____

3. $\dfrac{27}{10}$ = _____

4. $\dfrac{41}{8}$ = _____

5. $\dfrac{8}{3}$ = _____

1. $\dfrac{53}{7} =$ _____

2. $\dfrac{23}{7} =$ _____

3. $\dfrac{44}{6} =$ _____

4. $\dfrac{5}{2} =$ _____

5. $\dfrac{38}{8} =$ _____

1. $\dfrac{28}{8}$ = _____

2. $\dfrac{9}{4}$ = _____

3. $\dfrac{21}{9}$ = _____

4. $\dfrac{7}{2}$ = _____

5. $\dfrac{27}{4}$ = _____

1. $\dfrac{13}{6}$ = _____

2. $\dfrac{17}{3}$ = _____

3. $\dfrac{9}{4}$ = _____

4. $\dfrac{34}{5}$ = _____

5. $\dfrac{46}{7}$ = _____

Answers

Equivalent Fractions

Set 1

1. $\dfrac{5}{20} = \dfrac{6}{24} = \dfrac{2}{8} = \dfrac{1}{4} = \dfrac{3}{12} = \dfrac{4}{16} = \dfrac{7}{28}$

2. $\dfrac{6}{12} = \dfrac{1}{2} = \dfrac{3}{6} = \dfrac{2}{4} = \dfrac{7}{14} = \dfrac{4}{8} = \dfrac{5}{10}$

3. $\dfrac{6}{15} = \dfrac{8}{20} = \dfrac{12}{30} = \dfrac{4}{10} = \dfrac{2}{5} = \dfrac{14}{35} = \dfrac{10}{25}$

4. $\dfrac{6}{16} = \dfrac{18}{48} = \dfrac{15}{40} = \dfrac{21}{56} = \dfrac{12}{32} = \dfrac{9}{24} = \dfrac{3}{8}$

5. $\dfrac{6}{7} = \dfrac{36}{42} = \dfrac{42}{49} = \dfrac{30}{35} = \dfrac{12}{14} = \dfrac{18}{21} = \dfrac{24}{28}$

Set 2

1. $\dfrac{12}{18} = \dfrac{4}{6} = \dfrac{2}{3} = \dfrac{8}{12} = \dfrac{10}{15} = \dfrac{14}{21} = \dfrac{6}{9}$

2. $\dfrac{54}{60} = \dfrac{45}{50} = \dfrac{36}{40} = \dfrac{63}{70} = \dfrac{9}{10} = \dfrac{27}{30} = \dfrac{18}{20}$

3. $\dfrac{7}{42} = \dfrac{4}{24} = \dfrac{6}{36} = \dfrac{2}{12} = \dfrac{1}{6} = \dfrac{5}{30} = \dfrac{3}{18}$

4. $\dfrac{4}{9} = \dfrac{24}{54} = \dfrac{20}{45} = \dfrac{28}{63} = \dfrac{16}{36} = \dfrac{8}{18} = \dfrac{12}{27}$

5. $\dfrac{7}{49} = \dfrac{1}{7} = \dfrac{3}{21} = \dfrac{2}{14} = \dfrac{4}{28} = \dfrac{6}{42} = \dfrac{5}{35}$

Set 3

1. $\dfrac{4}{9} = \dfrac{8}{18} = \dfrac{12}{27} = \dfrac{16}{36} = \dfrac{20}{45} = \dfrac{24}{54} = \dfrac{28}{63}$

2. $\dfrac{3}{4} = \dfrac{6}{8} = \dfrac{9}{12} = \dfrac{12}{16} = \dfrac{15}{20} = \dfrac{18}{24} = \dfrac{21}{28}$

3. $\dfrac{5}{8} = \dfrac{10}{16} = \dfrac{15}{24} = \dfrac{20}{32} = \dfrac{25}{40} = \dfrac{30}{48} = \dfrac{35}{56}$

4. $\dfrac{3}{5} = \dfrac{6}{10} = \dfrac{9}{15} = \dfrac{12}{20} = \dfrac{15}{25} = \dfrac{18}{30} = \dfrac{21}{35}$

5. $\dfrac{2}{7} = \dfrac{4}{14} = \dfrac{6}{21} = \dfrac{8}{28} = \dfrac{10}{35} = \dfrac{12}{42} = \dfrac{14}{49}$

Set 4

1. $\dfrac{2}{7} = \dfrac{4}{14} = \dfrac{6}{21} = \dfrac{8}{28} = \dfrac{10}{35} = \dfrac{12}{42} = \dfrac{14}{49}$

2. $\dfrac{1}{2} = \dfrac{2}{4} = \dfrac{3}{6} = \dfrac{4}{8} = \dfrac{5}{10} = \dfrac{6}{12} = \dfrac{7}{14}$

3. $\dfrac{5}{6} = \dfrac{10}{12} = \dfrac{15}{18} = \dfrac{20}{24} = \dfrac{25}{30} = \dfrac{30}{36} = \dfrac{35}{42}$

4. $\dfrac{1}{10} = \dfrac{2}{20} = \dfrac{3}{30} = \dfrac{4}{40} = \dfrac{5}{50} = \dfrac{6}{60} = \dfrac{7}{70}$

5. $\dfrac{1}{3} = \dfrac{2}{6} = \dfrac{3}{9} = \dfrac{4}{12} = \dfrac{5}{15} = \dfrac{6}{18} = \dfrac{7}{21}$

Set 5

1. $\dfrac{4}{9} = \dfrac{8}{18} = \dfrac{12}{27} = \dfrac{16}{36} = \dfrac{20}{45} = \dfrac{24}{54} = \dfrac{28}{63}$

2. $\dfrac{1}{3} = \dfrac{2}{6} = \dfrac{3}{9} = \dfrac{4}{12} = \dfrac{5}{15} = \dfrac{6}{18} = \dfrac{7}{21}$

3. $\dfrac{4}{5} = \dfrac{8}{10} = \dfrac{12}{15} = \dfrac{16}{20} = \dfrac{20}{25} = \dfrac{24}{30} = \dfrac{28}{35}$

4. $\dfrac{1}{7} = \dfrac{2}{14} = \dfrac{3}{21} = \dfrac{4}{28} = \dfrac{5}{35} = \dfrac{6}{42} = \dfrac{7}{49}$

5. $\dfrac{1}{2} = \dfrac{2}{4} = \dfrac{3}{6} = \dfrac{4}{8} = \dfrac{5}{10} = \dfrac{6}{12} = \dfrac{7}{14}$

Set 6

1. $\dfrac{7}{10} = \dfrac{14}{20} = \dfrac{21}{30} = \dfrac{28}{40} = \dfrac{35}{50} = \dfrac{42}{60} = \dfrac{49}{70}$

2. $\dfrac{1}{6} = \dfrac{2}{12} = \dfrac{3}{18} = \dfrac{4}{24} = \dfrac{5}{30} = \dfrac{6}{36} = \dfrac{7}{42}$

3. $\dfrac{3}{4} = \dfrac{6}{8} = \dfrac{9}{12} = \dfrac{12}{16} = \dfrac{15}{20} = \dfrac{18}{24} = \dfrac{21}{28}$

4. $\dfrac{4}{7} = \dfrac{8}{14} = \dfrac{12}{21} = \dfrac{16}{28} = \dfrac{20}{35} = \dfrac{24}{42} = \dfrac{28}{49}$

5. $\dfrac{1}{8} = \dfrac{2}{16} = \dfrac{3}{24} = \dfrac{4}{32} = \dfrac{5}{40} = \dfrac{6}{48} = \dfrac{7}{56}$

Equivalent Fraction Problems

Set 1

1. $\dfrac{1}{2} = \dfrac{6}{12}$

2. $\dfrac{6}{15} = \dfrac{2}{5}$

3. $\dfrac{2}{3} = \dfrac{4}{6}$

4. $\dfrac{2}{4} = \dfrac{8}{16}$

5. $\dfrac{15}{18} = \dfrac{5}{6}$

Set 2

1. $\dfrac{6}{10} = \dfrac{3}{5}$

2. $\dfrac{24}{30} = \dfrac{4}{5}$

3. $\dfrac{20}{25} = \dfrac{4}{5}$

4. $\dfrac{5}{10} = \dfrac{1}{2}$

5. $\dfrac{3}{4} = \dfrac{12}{16}$

Set 3

1. $\dfrac{1}{5} = \dfrac{5}{25}$

2. $\dfrac{2}{4} = \dfrac{1}{2}$

3. $\dfrac{6}{24} = \dfrac{1}{4}$

4. $\dfrac{5}{30} = \dfrac{1}{6}$

5. $\dfrac{5}{6} = \dfrac{10}{12}$

Set 4

1. $\dfrac{5}{10} = \dfrac{1}{2}$

2. $\dfrac{4}{8} = \dfrac{2}{4}$

3. $\dfrac{1}{5} = \dfrac{2}{10}$

4. $\dfrac{4}{5} = \dfrac{12}{15}$

5. $\dfrac{6}{10} = \dfrac{3}{5}$

Set 5

1. $\dfrac{1}{2} = \dfrac{4}{8}$

2. $\dfrac{15}{20} = \dfrac{3}{4}$

3. $\dfrac{18}{24} = \dfrac{3}{4}$

4. $\dfrac{18}{36} = \dfrac{3}{6}$

5. $\dfrac{4}{6} = \dfrac{2}{3}$

Set 6

1. $\dfrac{18}{24} = \dfrac{3}{4}$

2. $\dfrac{1}{5} = \dfrac{5}{25}$

3. $\dfrac{2}{8} = \dfrac{1}{4}$

4. $\dfrac{6}{12} = \dfrac{1}{2}$

5. $\dfrac{18}{36} = \dfrac{3}{6}$

Mixed to Improper Number

Set 1

1. $9\frac{3}{4} = \frac{39}{4}$

2. $7\frac{2}{7} = \frac{51}{7}$

3. $7\frac{2}{5} = \frac{37}{5}$

4. $8\frac{2}{3} = \frac{26}{3}$

5. $7\frac{1}{2} = \frac{15}{2}$

Set 2

1. $7\frac{7}{10} = \frac{77}{10}$

2. $5\frac{1}{2} = \frac{11}{2}$

3. $7\frac{1}{3} = \frac{22}{3}$

4. $6\frac{5}{9} = \frac{59}{9}$

5. $7\frac{4}{5} = \frac{39}{5}$

Set 3

1. $2\frac{1}{8} = \frac{17}{8}$

2. $4\frac{5}{6} = \frac{29}{6}$

3. $8\frac{1}{3} = \frac{25}{3}$

4. $7\frac{5}{7} = \frac{54}{7}$

5. $9\frac{7}{9} = \frac{88}{9}$

Set 4

1. $8\frac{1}{4} = \frac{33}{4}$

2. $7\frac{2}{5} = \frac{37}{5}$

3. $2\frac{3}{7} = \frac{17}{7}$

4. $7\frac{1}{5} = \frac{36}{5}$

5. $6\frac{1}{2} = \frac{13}{2}$

Set 5

1. $4\frac{1}{5} = \frac{21}{5}$

2. $4\frac{1}{2} = \frac{9}{2}$

3. $2\frac{3}{4} = \frac{11}{4}$

4. $5\frac{2}{3} = \frac{17}{3}$

5. $9\frac{4}{5} = \frac{49}{5}$

Set 6

1. $8\frac{1}{4} = \frac{33}{4}$

2. $7\frac{2}{5} = \frac{37}{5}$

3. $2\frac{3}{7} = \frac{17}{7}$

4. $7\frac{1}{5} = \frac{36}{5}$

5. $6\frac{1}{2} = \frac{13}{2}$

Set 7

1. $3\frac{5}{8} = \frac{29}{8}$

2. $9\frac{1}{2} = \frac{19}{2}$

3. $3\frac{1}{2} = \frac{7}{2}$

4. $4\frac{6}{7} = \frac{34}{7}$

5. $9\frac{1}{5} = \frac{46}{5}$

Improper to Mixed Number

Set 1

1. $\dfrac{27}{4} = 6\dfrac{3}{4}$

2. $\dfrac{46}{8} = 5\dfrac{3}{4}$

3. $\dfrac{45}{10} = 4\dfrac{1}{2}$

4. $\dfrac{39}{10} = 3\dfrac{9}{10}$

5. $\dfrac{32}{7} = 4\dfrac{4}{7}$

Set 2

1. $\dfrac{21}{9} = 2\dfrac{1}{3}$

2. $\dfrac{13}{3} = 4\dfrac{1}{3}$

3. $\dfrac{37}{8} = 4\dfrac{5}{8}$

4. $\dfrac{13}{2} = 6\dfrac{1}{2}$

5. $\dfrac{29}{4} = 7\dfrac{1}{4}$

Set 3

1. $\dfrac{31}{4} = 7\dfrac{3}{4}$

2. $\dfrac{14}{4} = 3\dfrac{1}{2}$

3. $\dfrac{12}{5} = 2\dfrac{2}{5}$

4. $\dfrac{5}{2} = 2\dfrac{1}{2}$

5. $\dfrac{14}{5} = 2\dfrac{4}{5}$

Set 4

1. $\dfrac{10}{3} = 3\dfrac{1}{3}$

2. $\dfrac{38}{5} = 7\dfrac{3}{5}$

3. $\dfrac{27}{10} = 2\dfrac{7}{10}$

4. $\dfrac{41}{8} = 5\dfrac{1}{8}$

5. $\dfrac{8}{3} = 2\dfrac{2}{3}$

Set 5

1. $\dfrac{53}{7} = 7\dfrac{4}{7}$

2. $\dfrac{23}{7} = 3\dfrac{2}{7}$

3. $\dfrac{44}{6} = 7\dfrac{1}{3}$

4. $\dfrac{5}{2} = 2\dfrac{1}{2}$

5. $\dfrac{38}{8} = 4\dfrac{3}{4}$

Set 6

1. $\dfrac{28}{8} = 3\dfrac{1}{2}$

2. $\dfrac{9}{4} = 2\dfrac{1}{4}$

3. $\dfrac{21}{9} = 2\dfrac{1}{3}$

4. $\dfrac{7}{2} = 3\dfrac{1}{2}$

5. $\dfrac{27}{4} = 6\dfrac{3}{4}$

Set 7

1. $\dfrac{13}{6} = 2\dfrac{1}{6}$

2. $\dfrac{17}{3} = 5\dfrac{2}{3}$

3. $\dfrac{9}{4} = 2\dfrac{1}{4}$

4. $\dfrac{34}{5} = 6\dfrac{4}{5}$

5. $\dfrac{46}{7} = 6\dfrac{4}{7}$

39455743R00024

Made in the USA
San Bernardino, CA
26 September 2016